Map Skills

for Common Entrance Geography
and Key Stage 3

John Widdowson

HODDER
EDUCATION
AN HACHETTE UK COMPANY

Acknowledgements

Photo acknowledgements

The author and publisher would like to thank Paul Baker, Fiona Langridge and Simon Lewis for their helpful advice.

Cover & p.i John Townson/Creation; **p.2** Denny Rowland/Alamy; **p.3** John Widdowson; **p.4** *tl & bl* John Widdowson, *r* Collections/David Askham; **p.6** © Adrian Warren/Dae Sasitorn www.lastrefuge.co.uk; **p.12** John Widdowson, *inset* Collections/Liz Stares; **p.14** Collections/Paul Beko; **p.18** Skyscan; **p.21** Skyscan/© Highsight; **p.22** *photo* Simon Derry; **p.27** *t & b* John Widdowson; **p.29** © John Cleare/Mountain Camera Picture Library; **p.31** Collections/Ashley Cooper; **p.32** City & County of Swansea: Swansea Museum; **p.33** Collections/Ken Price; **p.35** *t* © Photolibrary Wales, *b* Collections/Ken Price; **p.36** Patrice Aguilar/Alamy; **p.38** NASA/USGS; **p.39** *t* © Vittoriano Rastelli/Corbis, *b* CuboImages srl/Alamy; **p.40** *t* © Mike Williams/Peak Pictures, *b* Chris Howes/Wild Places Photography/Alamy; **p.42** © Mike Williams/Peak Pictures.

t = top, *b* = bottom, *l* = left, *r* = right, *c* = centre

This product includes mapping data licensed from Ordnance Survey with the permission of the Controller of Her Majesty's Stationery Office. © Crown copyright. All rights reserved. Licence no. 100019872.

Answers to the questions and worksheet templates are available to download from www.hoddersamplepages.co.uk

Although every effort was made to ensure that the website addresses were correct at the time of going to press, Hodder Education cannot be held responsible for the content of any website mentioned in this book.

Papers used in this book are natural, renewable and recyclable products. They are made from wood grown in sustainable forests. The logging and manufacturing processes conform to the environmental regulations of the country of origin.

Orders: please contact Bookpoint Ltd, 130 Milton Park, Abingdon, Oxon OX14 4SB. Telephone: +44 (0)1235 827720. Fax: +44 (0)1235 400454. Lines are open from 9.00a.m. to 5.00p.m. Monday to Saturday, with a 24-hour message answering service. You can also visit our websites www.hoddereducation.co.uk and www.hoddersamplepages.co.uk

© John Widdowson 2005

First published in 2005
by Hodder Education,
an Hachette UK Company.
338 Euston Road
London NW1 3BH

Impression number   10
Year   2014

Layouts by Fiona Webb
Illustrations by Oxford Designers and Illustrators
Cover design by John Townson/Creation
Typeset in 11.5/13pt Sabon
Printed and bound in Dubai

A catalogue entry for this title is available from the British Library

ISBN: 978 0 340 90502 9

Contents

Acknowledgements   ii

Section 1

MAP SKILLS

1.1   What's special about maps?   2
1.2   Symbols – the key to using a map   4
1.3   A sense of direction   6
1.4   Finding places on a map   8
1.5   Distance – the long and short of it   10
1.6   Relief – the third dimension   12
1.7   A route with a view   14
1.8   Sketch maps and cross-sections   16
1.9   Decision time!   18

Section 2

STUDYING MAPS

2.1   Geomorphological processes
     • What is the flood risk in Shrewsbury?   20
     • Flooding in Shrewsbury, Autumn 2000   22
2.2   Settlements
     • What patterns do you find in a town?   24
     • How has Ashford grown?   26
2.3   Weather and climate
     • What happens to all that water?   28
     • Would you be safe on the mountain?   30
2.4   Economic activities
     • How has the Swansea Valley changed?   32
     • Was it a good location for new industry?   34
2.5   Tectonic processes
     • Is it safe to live in Zafferana?   36
     • Why live so close to a volcano?   38
2.6   Environmental issues
     • What can you do in a national park?   40
     • How should the park be managed?   42

Key to Ordnance Survey 1:50,000 maps   44

Index   46

1.1 What's special about maps?

You can't get far in geography without a **map**. Maps give us a **bird's-eye view** of the world, showing what it looks like from the air. Of course, these days with aeroplanes and satellites, it's possible for *everybody* to see the world from the air.

The photo on this page is an **aerial photo**. It shows the same area as the **Ordnance Survey (OS) map** on the map sheet. This is the map you will use in Section 1 (pages 2 to 19). In Section 2, you will use other OS maps.

I can't see where I'm going with this map in front of me.

Just look down dear. It's much easier!

A | Aerial photo of Seaford and Cuckmere Haven

A map of Britain. It shows where the photograph was taken.

Activities

1 To get the idea of maps, start by drawing some simple plans.
   a) Make a large copy of the table below.
   b) Draw a plan, or bird's-eye view, of each object in the table. One is done for you.

Object	Plan

2 Compare the photo on page 2 with the OS map. Then answer these questions.
   a) What is the name of the town?
   b) What is the name of the river?
   c) What colour is used to show the forest on the map?
   d) What colour is used to show the main road on the map?

3 a) What can you see on the photo on page 2 that is not on the map? Mention at least one thing.
   b) What can you see on the map that is not in the photo? Mention at least two things.

4 Look at this photo. It was taken somewhere on the map.
   a) Find the place on the map. What is it called?
   b) How is this photo different from photo A? (Clue: think about where the photo was taken from.)

5 You are planning a trip to Seaford. Which do you think would be more helpful in planning your trip – the OS map of Seaford or the aerial photo on page 2? Write one or two sentences to explain why.

Mysteries on the map

On the OS map you can see lots of lines. But what do they tell us?

There are straight blue lines, going from top to bottom and side to side. They are numbered at the edge of the map. Can you guess what they are for?

There are curvy brown lines with numbers on them that go almost everywhere. What do you think they show?

If you can't guess, you'll find out later in this section.

1.2   Symbols – the key to using a map

Maps are full of information, and most of it is shown in the form of **symbols** (the information wouldn't fit on the map if you had to write it all). Symbols can be small drawings, lines, letters, shortened words or coloured areas on the map. Look at the photos on this page. The features you can see are represented on Ordnance Survey maps by the symbols shown in the corner of the photos.

To understand the symbols on a map you need a **key**. The OS map on the map sheet has a key printed next to it. The key will help you to work out what all the symbols mean.

Activities

1 a) Name the three features in the photos in a table like this. Draw the symbol for each feature in the next column, using the correct colour. One is done for you.

Feature	Symbol
Church with a spire	✝

b) Add these features to your table. Find the symbols on the OS map key of Seaford and draw them in the second column of the table.

main road   youth hostel   golf course
footpath   single track railway   motorway
building   post office   view point   school

c) Now try to find all these features on your OS map. One is not on this map. Can you find which one?

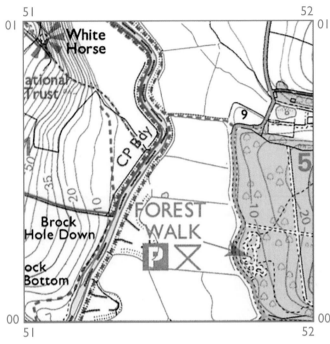

A Part of Seaford, enlarged from the map. This is an **urban** area. Most of the symbols here show **human** features, e.g. buildings.

B A section of the Cuckmere River valley from the map. This is a **rural** area. Most of the symbols here show **physical** features, e.g. a river.

2 Look at extracts A and B from the map and find the symbols in the OS map key. What do the symbols mean?
   a) List the urban and rural features that you find in each area. You should find at least six features in each area.
   b) Write a short paragraph to compare an urban area and rural area on an OS map. You could start like this:

   The urban area on the map has lots of human/natural features, but the rural area has…

3 Think of some map symbols of your own for features that are not shown in the key. (For example, a football stadium.) Think of at least five. Draw your own symbols for them.

4 Here is an island. It's a bit boring without any symbols on it!

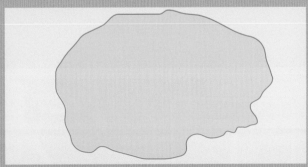

   a) Draw a large copy of the island in your book.
   b) Add symbols to show features on the island. Keep the symbols small, like they are on an OS map. You can be as imaginative as you want but don't make it too overcrowded.
   c) Give names to the towns and other important places on your map.
   d) Give the map a title and a key.

1.3   A sense of direction

Maps are useful if you want to give directions. But, first you need to know the points on the **compass** (A). Most maps, including all Ordnance Survey maps, have north at the top (it's always worth checking the north arrow on your map to be sure). Once you know where north is you just need to remember the order of the points on the compass. It's easy: go round clockwise from the top – Never Eat Shredded Wheat!

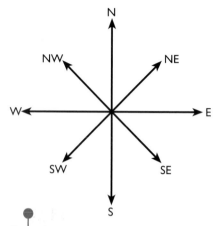

A | Main points of the compass

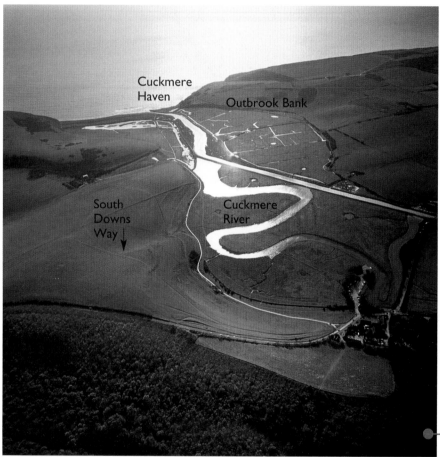

B   The Cuckmere Valley

Activities

1 Look at the letters on the grid below. Give
   directions for each of the following journeys.
   a) A to B   **b)** B to C   **c)** C to D
   d) D to E   **e)** E to A

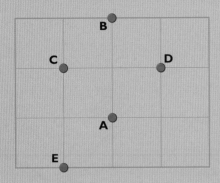

2 Follow the journey on the South Downs Way (a
   long distance footpath) from A to K, using map C.
   a) Give an accurate direction for each section of
      the journey, using the 16-point compass to
      help you. For example, 'A to B is NNW'.
      Carry on.
   b) Now, do the same, going in the opposite
      direction, from K to A.

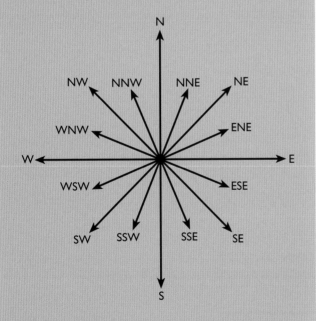

3 Look at photo B. Compare the photo with map
   extract C.
   a) In which direction was the camera pointing?
   b) How can you tell?

C   The South Downs Way.

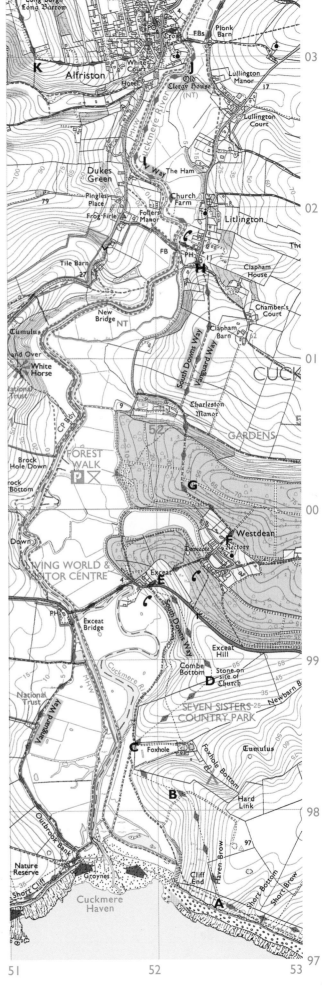

1.4 *Finding places on a map*

Looking for somewhere on a map can take ages if you don't know where to look. It's much easier if you can use a **grid reference**. Ordnance Survey maps use grid lines to divide the map into grid squares.

Find a four-figure grid reference

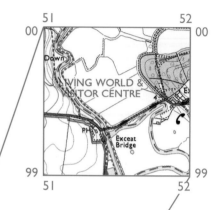

Look at the Seaford OS map. Find the **Living World & Visitor Centre** in square **5199**.

To find square **5199**, start at the bottom-left corner of the map. Go along the bottom of the map until you get to line **51**. Then go up the side of the map until you get to line **99**.

Follow the two lines until they meet at the **bottom-left** of square **5199**. This is a **four-figure grid reference**.

It's easy to give a four-figure grid reference. Always start in the bottom-left corner of the map. First, walk along the bottom – then, climb up.

Remember, don't climb before you can walk!

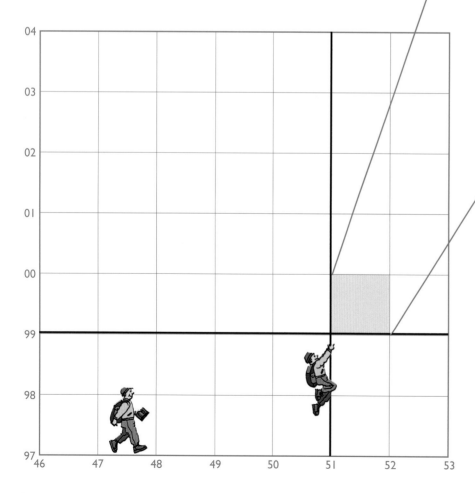

Find a six-figure grid reference

To be more accurate you can use a **six-figure grid reference** to pinpoint one place in your square. **The Living World & Visitor Centre** is at 518995.

Imagine the square is divided into a smaller grid with ten spaces across and ten up.

Start in the bottom-left corner of the square. Go across until you get to **8**.

Then go up the square until you get to **5**.

Follow the two imaginary lines until they meet at **518995**.

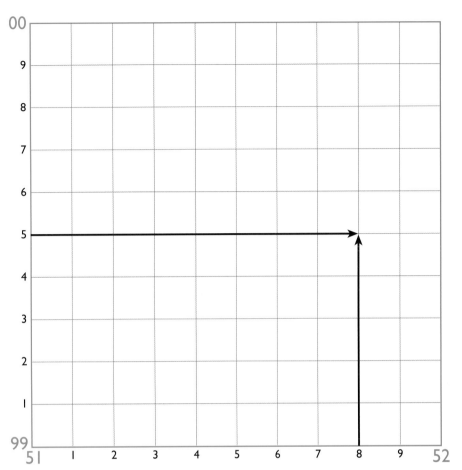

Activities

1 Look at the OS map. Match the following words on the map with the four-figure grid reference for the squares they are in.

> Seaford Bay   Seven Sisters Country Park
> Cuckmere Haven   Beacon Hill
> 5197   5298   4701   4699

2 Name the farms that you find in each of these squares on the map:
4602   5202   5099   4901.

3 Match the following features on the map with the correct six-figure grid reference.

> Exceat Bridge   White Horse
> Seaford Leisure Centre   Rathfinny Farm
> 497018   514994   493995   511009

4 Give six-figure grid references for these features on the map.

> Seaford Museum   Litlington Church
> Norton Top (a hill top)   Alfriston Post Office

5 The map tells a rather sad story. Write the story in your book, using the words that you find at the grid references on the map.

> 506973 did not feel very 481026. He had a 509011 temperature and his 529975 was sweating. After a while his face turned 494032 and he found a 498039 on his 464033. It was not very 462024. Soon it was all 512011. They wheeled him in a 513034 to the 497003.

1.5   *Distance – the long and short of it*

Distances on a map are always shorter than they would be in real life. To work out the actual distance you need to use the **scale**.

The scale on a map tells you how many times further the distance is in real life. There are three ways to show the scale:

- with words: 4 centimetres to 1 kilometre (one grid square). Therefore 1 centimetre equals 250 metres.

- as a ratio: 1:25,000 (you say 1 to 25,000)

- as a line:

```
0              1              2              3              4
|—————————————|—————————————|—————————————|—————————————|
                        Kilometres
```

The quickest way to measure distance on a map is to use the scale line (but, you can multiply by 25,000 if you prefer!).

Measure the straight-line distance (as the crow flies)

To measure the distance from Seaford Station (481992) to Exceat Bridge (514994):

Place the straight edge of a strip of paper on the map to join the two points. Mark the points on the paper.

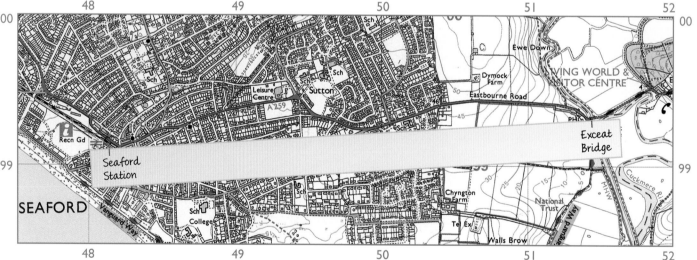

Transfer the paper to the scale line. Read off the real-life distance on the line. It is 3.3 km.

Scale 4 cm : 1 km

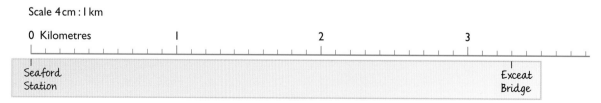

Measure the distance by road

Unfortunately, we can't fly like crows! Most of our journeys are by road. The most direct route from Seaford Station to Exceat Bridge is along the A259. The road is not straight, so this is how to measure it,

Lay the strip of paper on the first straight section of the journey, from the station to the junction near the leisure centre (494995). Mark the two points on the paper.

Where the road bends, swivel the paper so that it lays on the next section of road. Mark the next point (where the road bends again).

Repeat this each time the road bends, until you reach the bridge. Mark the last point on the paper.

Transfer the paper to the scale line. Read off the real-life distance between the first and last points on the paper.

The distance by road is 3.4 km.

Activities

1 Measure the distance as the crow flies from Seaford Station (481992) to the car park near the White Horse (509011). Read off the real distance to the nearest 100 m (0.1 km).

2 Now, measure the distance by road from Seaford Station to the White Horse car park. Start the journey on the main road and turn left at the junction near the leisure centre onto the secondary road (Alfriston Road).

3 Rachel goes to school in Seaford. Her school is at grid reference 487987.
   a) Each day her mum gives her a lift home in the car. Follow the route to work out which village Rachel lives in.

   > Turn right from the school and drive 1.2 km to the junction with the A259. Turn right again and drive for 2 km. Then turn left and drive 3 km north to the village.

   b) Sometimes her mum can't pick her up from school so Rachel has to go on the bus. The bus route goes through Alfriston and along the Alfriston Road. Work out the best place for Rachel to get off the bus. Give a grid reference. Now, measure the distance she has to walk home. She lives next to the pub in the village.

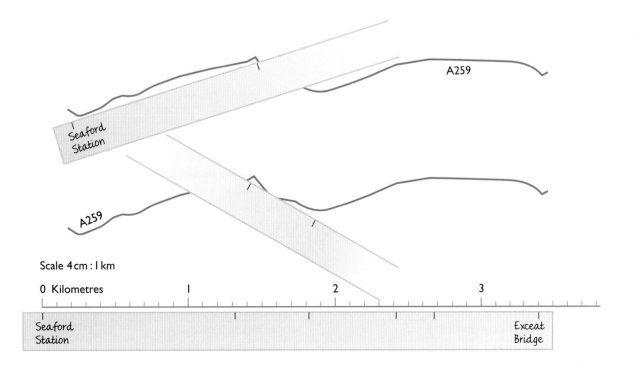

Scale 4 cm : 1 km

0 Kilometres   1   2   3

Seaford Station   A259   Exceat Bridge

1.6   Relief – the third dimension

Maps are really clever. They can turn a flat, two-dimensional piece of paper into a 3-D landscape (showing what the real shape is). The word we use in geography to describe the shape of the land is **relief**. Ordnance Survey maps use two methods to show relief.

- **Spot heights** They give the exact height for a spot on the map in metres above sea level. Sometimes hilltops are marked by a **triangulation pillar** – a small blue triangle on the map (see 526978).
- **Contours** These are thin brown lines joining places with the same height on the map. The numbers give the height in metres above sea level. The interval between the contour lines on this map is five metres.

There are no contour lines on hills in real life. You only see them on maps. Triangulation pillars are real. Here is one on a hilltop.

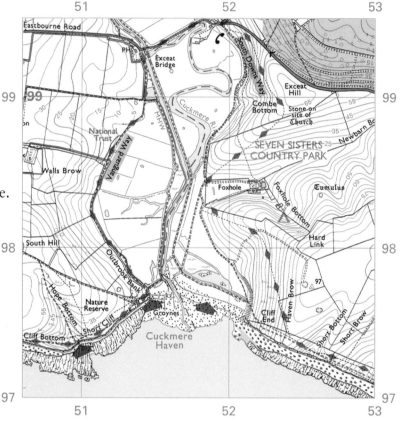

Activities

1 a) Find the spot heights at these grid references on your copy of the OS map. Give the heights.

> 498019   525977   467035   517006

   b) Find the highest spot height on the map. Give a grid reference.

2 Find the following journeys on the OS map. For each one, use contour lines to say if the journey is uphill, downhill or flat.
   a) By road from the car park (509011) to the Tile Barn (515015)
   b) By road from Exceat Bridge (514994) to Dymock Farm (505996)
   c) By footpath from Seaford Head (494977) to Cliff Bottom (510973)

   d) By footpath from New Bridge (516014) to Long Bridge (524036)

3 Find square 5100 on the OS map. The Cuckmere River flows from north to south. You are going to draw a simple sketch map to show the relief.
   a) Draw a large square, like this, in your book. Draw the Cuckmere River in the square.
   b) Draw the contour lines that you can see in the square, and number them.
   c) Now, label the sketch map. Label: a steep slope, a gentle slope, a flat valley bottom.

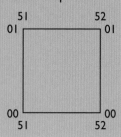

With a bit of experience you will learn to look at a map and recognise landscape features straight away. Here are some to get you started. See if you can find them on your OS map sheet.

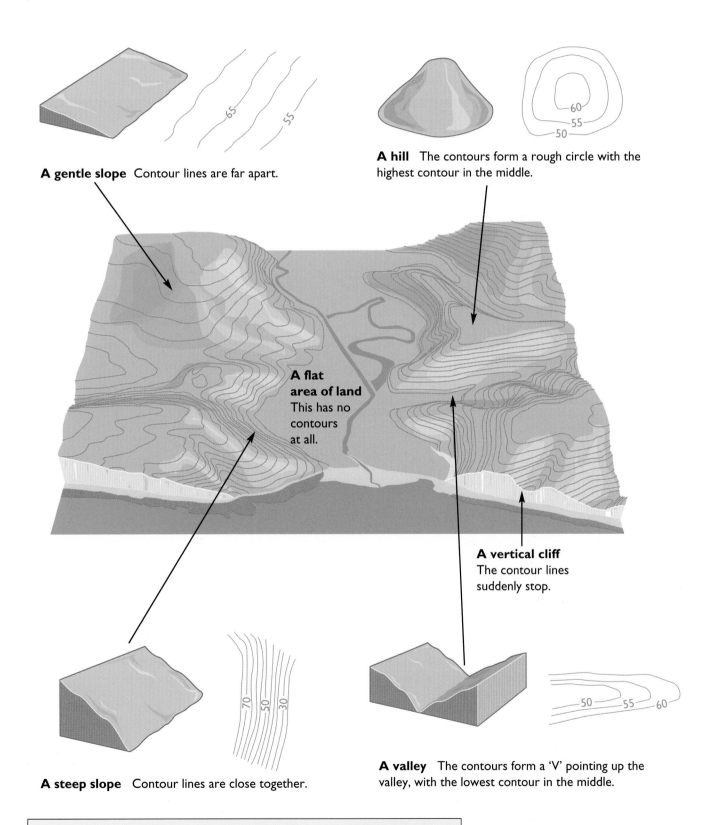

A gentle slope Contour lines are far apart.

A hill The contours form a rough circle with the highest contour in the middle.

A flat area of land This has no contours at all.

A vertical cliff The contour lines suddenly stop.

A steep slope Contour lines are close together.

A valley The contours form a 'V' pointing up the valley, with the lowest contour in the middle.

A simple rule to remember is: the closer the contours the steeper the land.

1.7   A route with a view

Look at this spectacular view. It shows the Seven Sisters chalk cliffs on the south coast near Seaford, in square 5297 on the Ordnance Survey map extract (you can only see two sisters on the map). Can you work out why they are called the *Seven* Sisters?

A | The Seven Sisters

You don't need a photograph to know what the view would be like – a map is good enough. You can interpret the symbols and contour lines on the map to imagine the view.

Activities

I   Look at photo A. Compare it with this labelled square from the OS map.

   a) Which is the easiest feature to recognise in the photo?
   Why is it harder to recognise on the map?

   b) Now draw a sketch of photo A. Label the same features on your sketch.

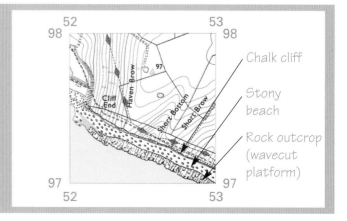

Chalk cliff

Stony beach

Rock outcrop (wavecut platform)

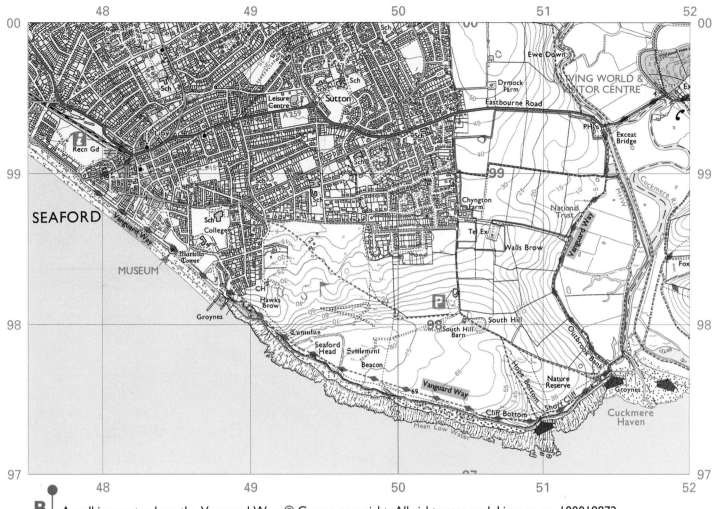

2 You go for a walk from Seaford Station (481992) to Exceat Bridge (514994) along Vanguard Way. The route is shown on map extract B.

a) Divide the walk into sections between these grid references;

481992   478989   488982   494977
510973   514977   514994

b) Complete a large table, like the one below, to describe each section of the walk, using information on the map. Mention grid references, distance, direction, slope (up, down or flat), and interesting features that you see. The first section of the walk is done for you.

3 Use the table you completed in question 2 to write an account of your walk. Write a paragraph to describe each section. The first paragraph could be:

We came out of Seaford Station (481992) and turned right onto the Vanguard Way. We walked 400 metres south-west towards the seafront. The ground was flat.

4 At which point on the walk would you see the view in photo A?

a) Give a six-figure grid reference for the point where you are standing.

b) In which direction are you looking?

From	Grid reference	To	Grid reference	Distance	Direction	Slope	Interesting features
Seaford Station	481992	Seafront	478989	0.4 km (400 m)	SW	Flat	Stony beach

15

1.8 *Sketch maps and cross-sections*

Ordnance Survey maps give lots of detail. Sometimes, you don't need all that information – **a sketch map** will do. Drawing a sketch map is one of the important skills you learn in geography.

This extract from the map shows the area around the village of Alfriston. Imagine that you wanted to draw a sketch map to show the site of the village...

Here is the sketch map that you might draw.

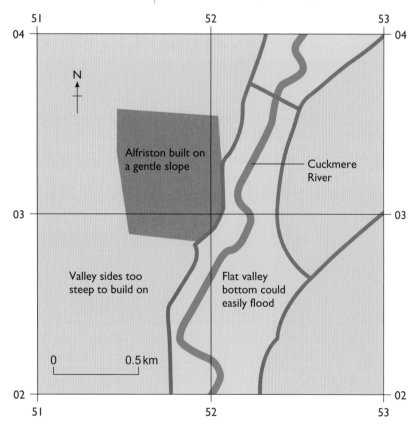

Sketch map to show the site of Alfriston

Draw a sketch map

- Draw a frame for the map. You can make it as large or small as you need to.
- Divide the frame into grid squares like the map. They will act as a guide to draw on.
- Draw the main features from the map, using a pencil. Don't include unnecessary details, like buildings or contour lines.
- Colour or shade the map to highlight the features.
- Give the map a title, north arrow and scale. Use a key if you need to.
- Label, or annotate, the map to explain what it shows.

Key

- River
- Road
- Flat land
- Steep slope
- Built-up area

Draw a cross-section

Another important geographical skill is drawing a **cross-section**. A cross-section is an imaginary slice through the landscape. It can help you to visualise what the relief is like.

- Place the straight edge of a strip of paper along the line of the section on the map.
- Mark the two ends of the section on the paper.
- Mark each point where a contour line touches the paper. Write the height of the contour line beside each mark. Mark other important features on the section, such as rivers.
- Draw a grid for the cross-section onto graph paper. The horizontal axis should be the width of the section. The vertical axis should go up to the highest contour line.
- Transfer the strip of paper to the horizontal axis. Now, mark each contour with a '✗' at the correct height above the axis.
- Join the marks with a pencil line to show the relief. Label any features on the line.

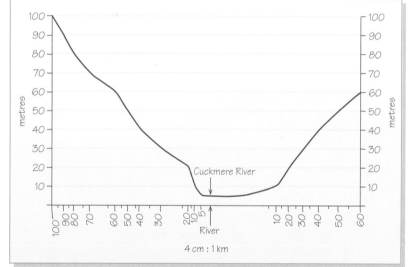

Activities

1 Draw a sketch map to show the route that you followed on the walk along the Vanguard Way that you planned in activity 2 on page 15. Draw a large frame and grid, like this, to start.

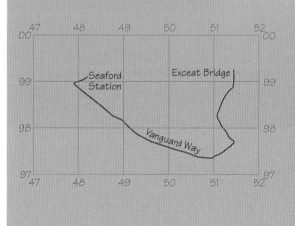

2 Make a copy of this grid (below) on graph paper. Then use it to draw a cross-section across the Cuckmere valley, along grid line 99, from grid reference 510990 to 530990 on OS map.

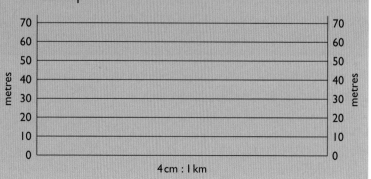

1.9 Decision time!

Maps are used when important decisions have to be made. You are going to imagine that you are the directors of a well-known supermarket company. The company wants to open a new superstore at Seaford. You have to look in and around the town for the best site.

Photo A shows how big a modern superstore can be. The company needs a lot of land to build the superstore and for the giant car park. You are looking for a site of about four hectares. That's about this size on the map (1:25,000 scale):

A | A modern superstore and car park

To meet the company's requirements, the site must also be:

- close to where people live, so they don't have to travel too far
- well connected by roads, so people can travel by car
- on flat land to make it easier to build
- on well-drained land, away from any rivers or streams that could flood.

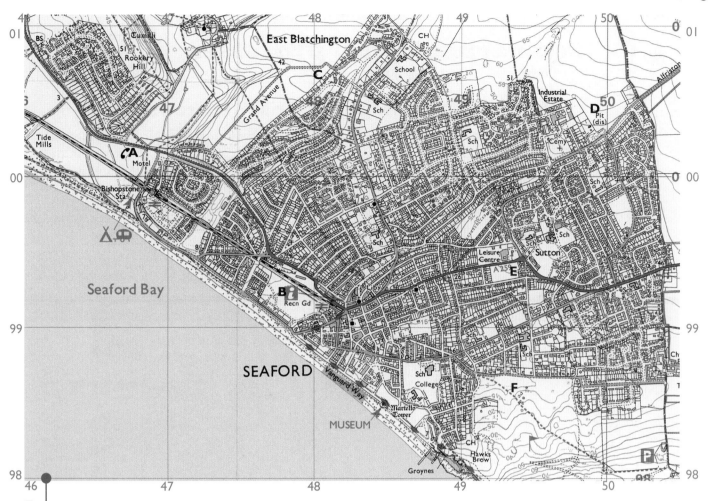

B Six possible sites for a superstore, in and around Seaford. © Crown copyright.
All rights reserved. Licence no. 100019872

Activities

1 Work with a small group. You are the directors of the supermarket company.

a) Find the six possible sites for a superstore on map extract B. Match the sites A to F with the correct grid references.

478992	494986	480007	467002
	499004	494994	

b) Cut out a small piece of paper, the size of the site that the company is looking for (see page 18). Place it on each possible site on the map.

c) Choose the best site for the new superstore based on the company's requirements.

d) Write a letter to the town council, asking for planning permission to build the new superstore. Tell them which site you chose. Give your reasons for choosing this site.

2 Now, with your group, change roles. You are going to represent the town council in Seaford.

You have received a planning application from a supermarket company to build a new superstore.

The town council have their own views about the development of new shopping facilities in Seaford. They aim to:

• provide a good range of shopping in Seaford
• ensure easy access to shops for everybody, including people without cars
• keep traffic congestion in the town to a minimum
• prevent building on open space that is important for recreation.

a) Swap planning applications with another group. Read their letter. Then, consider the site that they have chosen for a new supermarket. Decide whether you would give planning permission for it to be built there.

b) Write a reply to the company to give your decision and reasons for it.

2.1 *Geomorphological processes*

What is the flood risk in Shrewsbury?

The River Severn is the longest river in Britain. It flows from its source in the Cambrian Mountains of Wales to its estuary in the Bristol Channel. Many settlements have been built along the course of the river. One of these is the town of Shrewsbury.

In the past, people settled near rivers because they provided a good water supply, transport and protection. Unfortunately, there is also a downside to living near rivers – **floods.**

Look carefully at map A. It shows the River Severn flowing through Shrewsbury today. Why do you think the town was built here? And is there any risk of flooding?

A 1:25,000 OS map extract of Shrewsbury, on the River Severn.
© Crown copyright. All rights reserved. Licence no. 100019872

B | View of the River Severn flowing through Shrewsbury

Activities

1 Look at map A.
   a) What do you find at grid reference 494128? How does it help to explain why Shrewsbury was built here?
   b) How many bridges can you find crossing the River Severn in Shrewsbury? Give a six-figure grid reference for each one.
   c) Why was a bridging point on the river also a good place to build a settlement?

2 Look at photo B, and compare it with map A.
   a) Find the places labelled in the photo on the map. Give the four-figure grid reference for the square they are in.
   b) In which direction was the camera pointing when the photo was taken?

3 **a)** Draw a sketch of photo B. It is started for you on the right.

Label each of these features on your sketch:

> a meander   a bridge   Shrewsbury
> the castle   the football stadium
> an area of low flat land (flood plain)

   b) Are either the castle or football stadium at risk from flooding?
   c) Why were these two sites chosen, do you think?

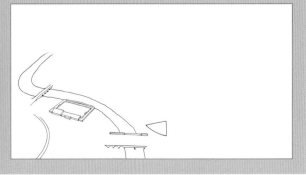

Flooding in Shrewsbury, Autumn 2000

In Autumn 2000, the worst floods for over fifty years devastated towns all along the River Severn. Shrewsbury was badly affected. It was flooded three times in the space of six weeks. No sooner had people finished mopping up after one flood, than they were up to their knees in water again!

One place in Shrewsbury that floods regularly is the football stadium. You saw it in photo B on page 21. It is on the low-lying **floodplain**, which is the natural overflow area for the river when it carries too much water. Unfortunately, as the town has grown over the years, other parts of the floodplain have been built on. These are the areas where the flood risk is greatest.

C Ganges Tandoori restaurant in Shrewsbury, flooded in Autumn 2000

D Floodplain map for Shrewsbury, showing the areas at risk from flooding (source: The Environment Agency)

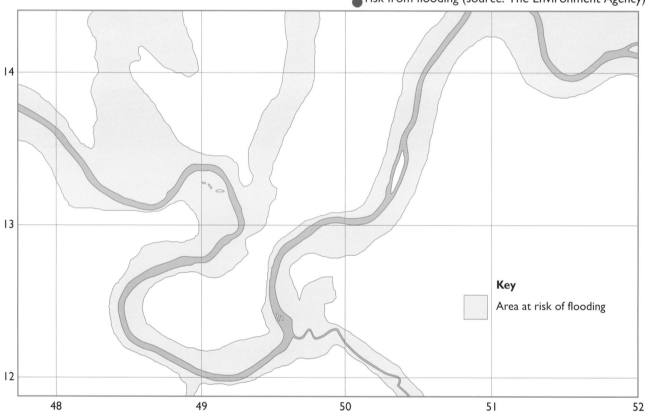

Key

Area at risk of flooding

Activities

1 Look at map D. Compare it with map A on page 20, to find the places that are at risk from flooding.
   a) Name the places that you find at each of these grid references on map A.

   | 491125 | 484134 | 487130 | 509144 |
   | 484122 | 491131 |

   b) In each case, find out if the place is at risk from flooding or not, from map D.
   c) What do you notice about the places that are at risk from flooding, and those that are not?
   d) Was it sensible to plan the town in this way, do you think? Explain why.

How can the flood risk be reduced?

In recent years, floods along the River Severn have been happening more often. Some people think this is because our climate is changing. Whatever the reason, something has to be done to reduce the risk.

After the floods in Autumn 2000, the **Environment Agency** designed a flood defence scheme to reduce the risk from future floods in Shrewsbury. They built a **floodwall** and **embankment** along sections of the river (see E). The aim was to protect areas of Shrewsbury that are prone to flooding.

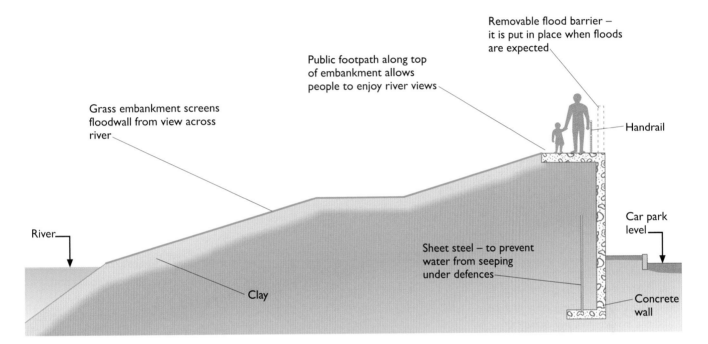

Removable flood barrier – it is put in place when floods are expected

Public footpath along top of embankment allows people to enjoy river views

Grass embankment screens floodwall from view across river

Handrail

River

Car park level

Sheet steel – to prevent water from seeping under defences

Clay

Concrete wall

E Cross-section of the design for a floodwall and embankment

Assignment

You work for the Environment Agency in Shrewsbury. You have been asked to decide where to build floodwalls and embankments along the river. Flood defences are very expensive, so you should only build where it is absolutely essential to protect the town.

1 Look at map A on page 20. Draw a sketch map to show the river, main roads, railway and built-up area. Mark and label important places on the map (e.g. castle).
2 Look at map D on page 22. Highlight the areas at risk from flooding on your map.

3 Decide which sections of the bank, on either side of the river, need flood defences to protect the town. Mark them on your map.
4 Write a short report to explain the decisions that you took.

STOP PRESS

When water in the River Severn in Shrewsbury rose to flood level again in February 2004, the new flood defences were used for the first time. The new floodwall and embankment did their job. There was no flooding in the town.

Find out more at these two websites:
www.environment-agency.gov.uk/regions/midlands
www.shrewsbury.gov.uk/public/leisure/gallery/flooding.htm

2.2   Settlements

What patterns do you find in a town?

Towns and cities can seem just like a jumble of buildings and roads when you are in the thick of them. But look at them on a map, and you might be able to see a pattern.

Here is an urban model to show the **land use pattern** that you find in a typical town or city (see A). Compare the model with the town of Ashford in map B. Can you see a pattern?

B   1:25,000 OS map extract of Ashford in Kent. © Crown copyright. All rights reserved. Licence no. 100019872

Central business district (CBD) is at the centre, where most shops and offices are. It is surrounded by an inner ring road. There are also restaurants, theatres, museums and other public buildings. Many of the buildings are large.

A | Urban land use model

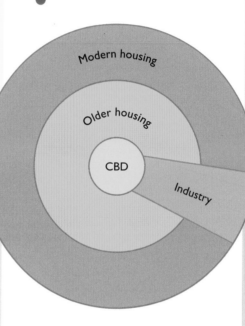

Older housing (pre-1970s) is usually found closer to the centre, around the CBD. The houses are in rows and the streets are straight.

Modern housing (since 1970) is usually built closer to the edge of the town. The houses are in groups and the streets are not straight. Houses and gardens are often smaller.

Industry is found around the main transport routes, both roads and railways. It includes factories, industrial estates and business parks. Most of the buildings are very large.

Activities

1 Find the given examples of these four land uses in map B opposite.

> CBD   older housing   modern housing   industry

Give four-figure grid references for the squares the examples are in.

2 Locate the CBD in squares 0042 and 0142 on map B opposite.

What evidence can you find that this is the CBD? Find at least four bits of evidence. Give a six-figure grid reference for each one.

3 Find the squares with industry in and around Ashford on map B.
   a) Give a four-figure grid reference for each one.
   b) Write one or two sentences to describe the distribution of industry on the map.
   c) Does the distribution of industry fit the urban land use model? Give reasons for your answer.

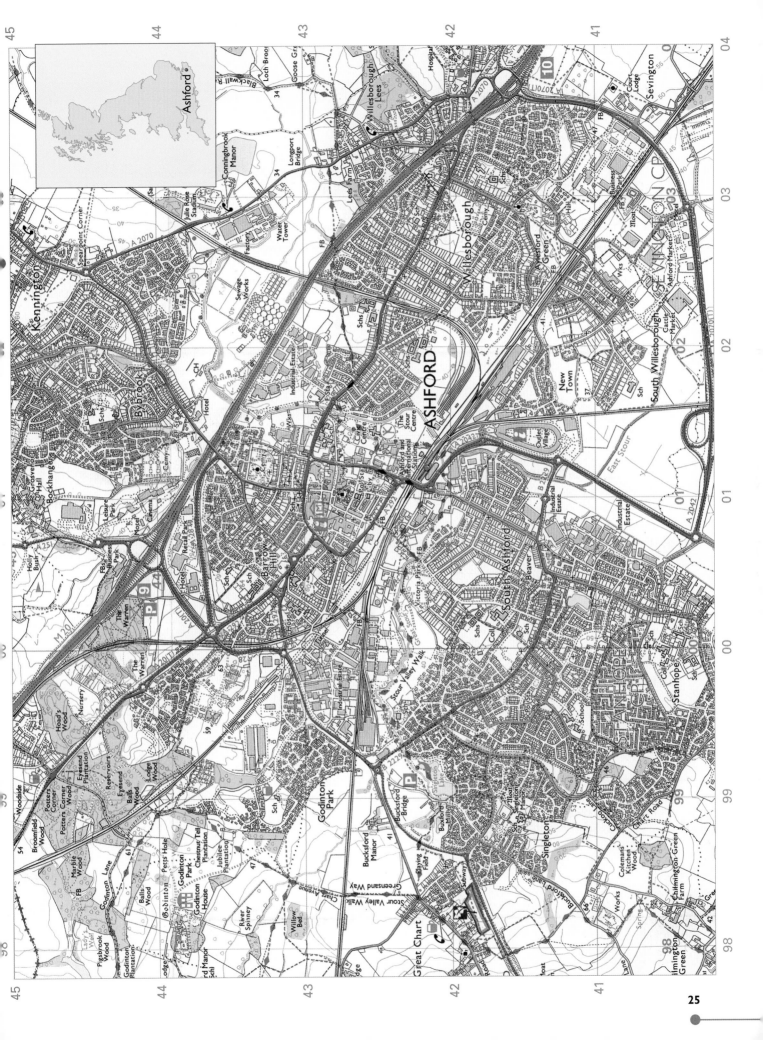

How has Ashford grown?

Ashford is one of the fastest growing towns in the UK. In 1970, its population was 79,000 but by 2005, it had grown to 105,000.

One reason for Ashford's growth is its location, on the doorstep of Europe. The town lies on the M20 motorway, between London and the Channel ports of Folkestone and Dover. It also has its own International Station with direct links to Paris and Brussels via the Channel Tunnel.

These transport connections have brought lots of industry and business to Ashford, creating new jobs – and jobs bring people.

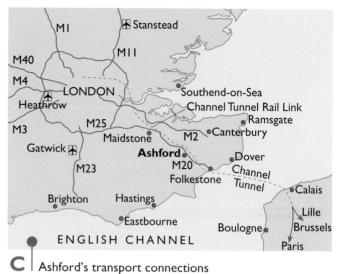

C Ashford's transport connections

Activities

1   You are going to draw a map to show how Ashford has grown.
    a) First, complete a large copy of this land use map, based on map B on page 25. Shade the CBD and industrial areas on the map. Draw a key.
    b) The remaining areas on the map show housing. Locate the areas on map B to find out if they are older housing or modern housing. Shade them on your map using two more colours. Complete the key.
    c) Modern housing was built after 1970. Draw a line around the older housing on the map to show the area of Ashford in 1970.

2   Use the map you have drawn to work out how much Ashford has grown. Find the area of the town:
    a) in 1970
    b) today.
    To do this count the number of grid squares on the map (only count squares that are over half full). Each square on the map is 1 km².

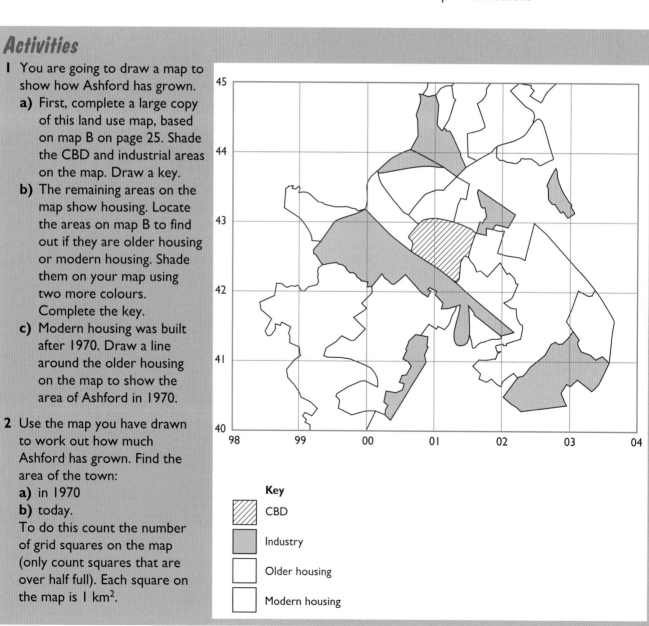

Key

- ▨ CBD
- ▧ Industry
- ☐ Older housing
- ☐ Modern housing

And how should it grow now?

The UK needs thousands of new homes. Most of them are needed in the south-east of England where population is growing fastest.

The government has chosen the areas where it wants most of the new homes to be built. One of them is Ashford. The population of the town is expected to rise to 140,000 by 2020. The question is – where should all the new homes be built?

There are two options. One is to build on **greenfield** sites in rural areas outside the town, like photo D. Lots of people want to live in the countryside these days, with more space around them. The other option is to build on **brownfield** sites in the town. These are sites that were built on before, but need to be redeveloped, like photo E. The government favours using brownfield sites to preserve the countryside. What do you think is the best option?

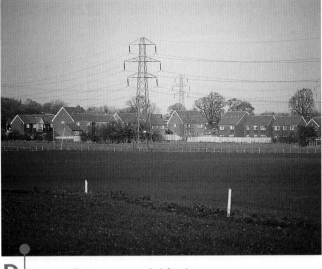

D A greenfield site near Ashford

E A brownfield site in Ashford

Assignment

A number of sites for new homes have been proposed in, and around, Ashford. Some are on greenfield sites and some on brownfield sites. Two of the proposed sites on map B on page 25 are:

Chart Road at 988406

Old railway works at 017416

a) Which is a greenfield site and which is a brownfield site? Explain your answer.

b) Look at the statements below. Which statement is most likely to be true for each site? In each case give a reason.

c) Choose the best site to build new homes. Write a paragraph to explain your choice.

| It has more pleasant surroundings |
| It is more peaceful |
| It would be easy to reach the town centre |
| It would make good use of wasteland |
| It would require more trees to be chopped down |
| It would be further from shops |
| It would lead to greater traffic congestion |
| It is more prone to flooding |
| It could be noisy |
| It would mean less land available for farming |
| It is quite hilly |
| It would be convenient to get to the station |

2.3   Weather and climate

What happens to all that water?

The Lake District is one of the wettest parts of the British Isles. During an average year it gets rain on 200 days out of 365. The total annual rainfall is about 2,000 mm, compared to just 600 mm in London. It is a good place to see the **water cycle** at work.

All that water has an impact on the landscape. Rivers and lakes abound, and the rainfall makes everywhere green (except when it is white in winter!). You can see the evidence all over map A.

A   1:50,000 OS map extract of Helvellyn, the third highest peak in the Lake District. (Notice that the scale of the map is 1:50,000, or 2 cm to 1 km. You can find a key for 1:50,000 OS maps on pages 44 to 45.) © Crown copyright. All rights reserved. Licence no. 100019872

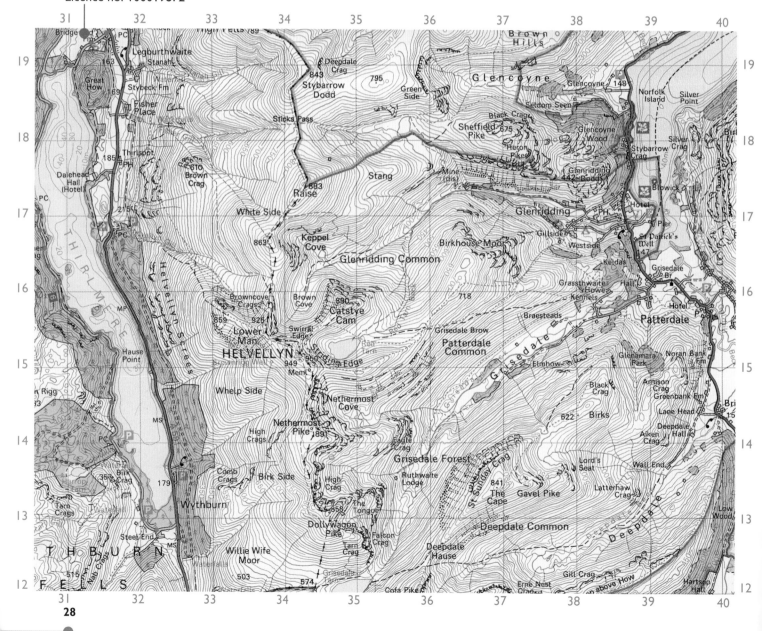

B | The view from Helvellyn looking across Thirlmere.

Activities

I Look at map A.
  a) Find Helvellyn. Give the six-figure grid reference for the summit. (Look for the triangulation pillar.)
  b) What is the height?

2 Complete the table below to describe evidence of the water cycle on map A.
  a) In the second column, name the feature that you find at each grid reference.
  b) In the third column, explain how the feature is connected with the water cycle.
  One is done for you.

Grid reference	Feature	Connection with the water cycle
347153	Red Tarn	Water is stored here
365146		
400131		
312163		
338151		

3 There are many rivers and lakes on the map. Does this tell you that the rock in this area is permeable or impermeable? Explain your answer.

4 Look at photo B. You are going to draw an annotated sketch of the view to describe the water cycle in the Lake District.
  a) Draw a sketch, starting like this, showing the main features in the photo.

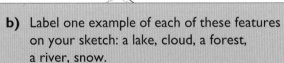

  b) Label one example of each of these features on your sketch: a lake, cloud, a forest, a river, snow.
  c) Now match the information below with the correct label. Write it beside the label to annotate your sketch.
   • water runs off the surface
   • water is transpired into the air
   • water vapour condenses into droplets
   • water is stored here and some evaporates
   • precipitation falls on the ground.

Would you be safe on the mountain?

Thousands of people go walking in the Lake District every year. To plan a walk safely in the mountains you need to be a good map reader, and a good weather forecaster!

Look at contour map C. It has been drawn from map A, by simplifying the contour lines. Instead of showing every contour at 10 metre intervals, like map A, it only shows contours every 50 metres.

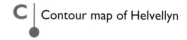

C | Contour map of Helvellyn

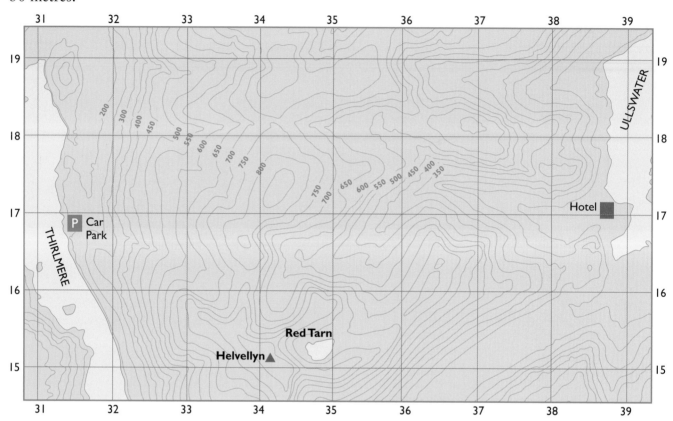

Activities

1 Look at contour map C. You are going to draw a cross-section along grid line 17 from the car park beside Thirlmere (315169) to the hotel in Glenridding (387171).

a) Place the straight edge of a strip of paper on grid line 17 on contour map C, between the car park and hotel. Mark the contours.

b) Draw a grid this size in your book to draw the cross-section on.

c) Transfer your strip of paper to the horizontal axis. Mark the contours at the correct height above the axis. Join the points to complete the cross-section.

2cm : 1 km

Assignment

D Along Browncove Crags from Helvellyn Lower Man

You are going to plan a walk to the top of Helvellyn, starting at the car park beside Thirlmere (315169) and ending at the hotel in Glenridding (387171). First, work out the best route on the map, then decide which day will have the best weather to do the walk.

1 Work out the best route for the walk, using map A.
   a) Follow the footpaths on the map from the car park up to the top of Helvellyn, and then down to the hotel. You could choose the shortest route or the route that avoids the steepest hills.
   b) Describe your route. You could start like this;

Leave the car park beside Thirlmere (315169) and walk up a steep hill for about 1.3 km heading south-east. The path follows Helvellyn Gill. Where the footpaths cross (327164) turn right and continue…

2 Now decide which day to do the walk, using the weather forecast information.
   a) Think about how the weather conditions could affect your walk. Here are four factors you need to consider:

   - rain makes the ground slippery and reduces visibility
   - low cloud reduces visibility and makes hilltops and ridges dangerous
   - strong wind makes hilltops and ridges dangerous (wind speed increases as you go up)
   - temperature falls as you go higher (1°C for every 100 metres you go up). Below 0°C snow falls and water on the ground freezes.

   b) Read the forecast for each day and decide whether it would be safe to do the walk that day, or not. Give reasons for your decision.
   c) If you had to choose one day to do the walk, which day would it be? Why?

WEATHER FORECAST

Date: 29.11.05

Sunrise: 8.02   **Sunset:** 16.10

General remarks: Depression moving in from the west. Rain approaching.

Wind: Strong westerly wind

Temperature: 11°C at 200 m (lake level)

Cloud: Cloud base at 500 m

WEATHER FORECAST

Date: 30.11.05

Sunrise: 8.03   **Sunset:** 16.09

General remarks: Low pressure moving away to the east. Heavy showers

Wind: Gusty, north-westerly wind

Temperature: 8°C at 200 m (lake level)

Cloud: Broken cloud. Cloud base at 1,000 m

WEATHER FORECAST

Date: 1.12.05

Sunrise: 8.04   **Sunset:** 16.08

General remarks: High pressure. Sunny

Wind: Moderate, northerly wind

Temperature: 5°C at 200 m (lake level)

Cloud: None

2.4 Economic activities

How has the Swansea Valley changed?

The Lower Swansea Valley in South Wales has a long industrial history. Coal was first mined here back in the Middle Ages. In the nineteenth century, **coal mining** grew rapidly as metal-smelting industries developed in the valley. Boats brought metal ore up the river from the docks to where the coal was mined from the valley sides.

For over 100 years the Swansea Valley was an ideal location for industry (photo A). But by 1952 – the year that map B was published – coal mining was in decline. Coal was getting more difficult and expensive to mine. By 1984 the last mine in the valley had closed and the metal smelting industry went too.

A The Lower Swansea Valley in the 1950s

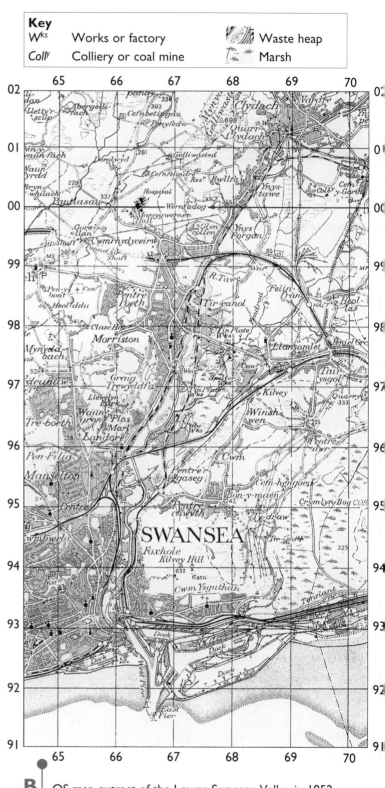

Key
W^{ks}   Works or factory
Coll^y   Colliery or coal mine
    Waste heap
    Marsh

B OS map extract of the Lower Swansea Valley in 1952 – the scale on the map is one inch to one mile (roughly 1.5 cm to 1 km, or 1:66,000). Reproduced from 1952 Ordnance Survey map with the kind permission of the Ordnance Survey

Today, the Swansea Valley is hardly recognisable from the grimy, polluted place it was 50 years ago. Gone are all the old industries and, in their place, new **industrial estates** have been built. **Tourism** is another recent economic activity in the valley.

D The Swansea Valley today

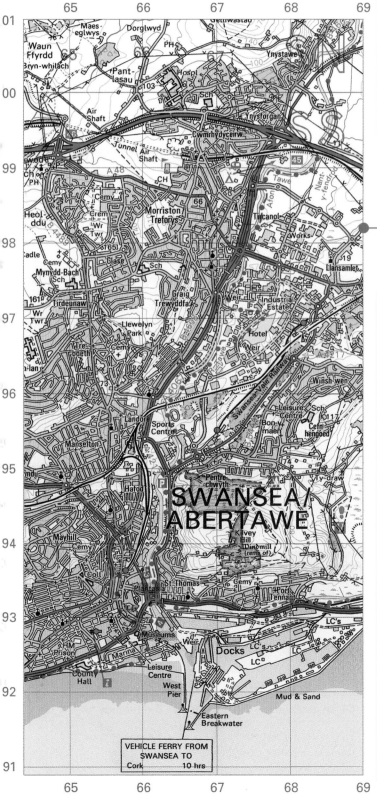

C 1:50,000 OS map extract of the Lower Swansea Valley in 2003

Activities

1 Look at map B.
   a) Find at least three examples of the metal industry in 1952 on the map. Give six-figure grid references.
   b) By 1952 there was only one coal mine left in the valley. Can you find it? Give a grid reference.

2 Look at map C. The key to this map is found on pages 44 to 45. Find at least six bits of evidence on the map to show the importance of the tourist and leisure industries in 2003. Give six-figure grid references. For example, leisure centre at 658926.

3 Compare transport in the Swansea Valley in 1952 and 2003 using the two maps. Answer these questions:
   a) In 1952, the docks were used for importing and exporting goods. How are the docks used in 2003?
   b) What type of transport did industry use in 1952? What transport do industrial estates use in 2003?
   c) What new transport route can you find on the 2003 map? Why is it important for industry and tourism?

Was it a good location for new industry – in 1952...

By the second half of the twentieth century, the metal-smelting industry in the Lower Swansea Valley was in decline. Many of the factors that brought factories there in the nineteenth century had changed (map E). In addition, industry itself was changing. It was hard to attract new industries to the area.

E   Sketch map of the Lower Swansea Valley in 1952, showing advantages and disadvantages for industry

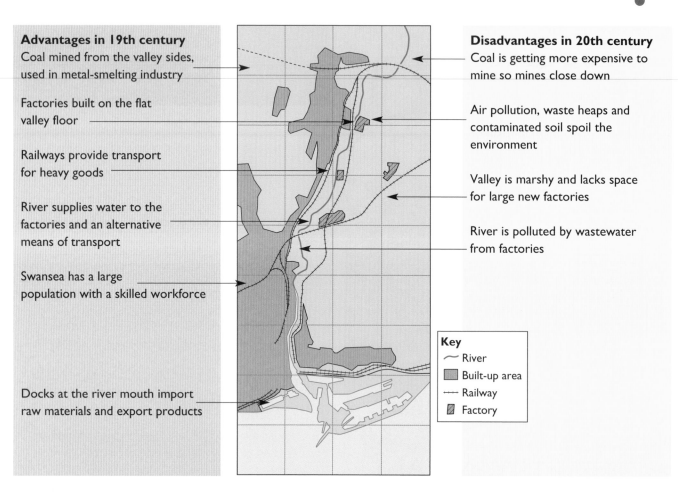

Advantages in 19th century

Coal mined from the valley sides, used in metal-smelting industry

Factories built on the flat valley floor

Railways provide transport for heavy goods

River supplies water to the factories and an alternative means of transport

Swansea has a large population with a skilled workforce

Docks at the river mouth import raw materials and export products

Disadvantages in 20th century

Coal is getting more expensive to mine so mines close down

Air pollution, waste heaps and contaminated soil spoil the environment

Valley is marshy and lacks space for large new factories

River is polluted by wastewater from factories

Key
- ～ River
- ▨ Built-up area
- ┉ Railway
- ▨ Factory

Activities

1 Look at map E. Compare it with photo A and map B on page 32.
   Complete a large table like this to list the advantages and disadvantages of the Lower Swansea Valley for industry. Give evidence from photo A and/or map B to support your statements. The first one is done for you.

2 You are the owner of a car manufacturing company in 1952. Would you choose the Swansea Valley to build a large new factory? Give your reasons.

Advantages in 19th century	Evidence	Disadvantages in 20th century	Evidence
Coal can be mined from the valley	Coal mine at 696003	Coal is getting expensive to mine so mines close down	Only one mine on the map

... and now?

The Lower Swansea Valley has been **regenerated** since 1980. Derelict buildings, waste heaps and contaminated soil have been removed. In their place, new industrial estates, leisure facilities, and even a lake, have been created. The old docks have been turned into a marina (photo F) and disused railway tracks are used as cycle routes (photo G). The valley has become an area where industry invests, and tourists want to visit.

F | Swansea Marina

G | Cycling on a disused railway track

Assignment

You are going to produce a tourist map for visitors to the Lower Swansea Valley today.

1 First, look at map C to identify all the features that might be of interest to visitors (you found some in Activity 2 on page 33).

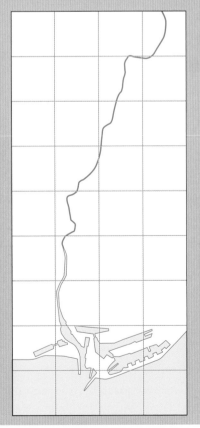

2 Draw a large sketch map of the valley, starting like the one shown here. Mark the tourist features on your map and put them on a key.

3 Add other important information to the map, including the main transport connections that visitors could use. Add them to the key.

4 Annotate the map to make the Swansea Valley sound like an exciting place to visit. For example: *You don't need to own a boat to enjoy the Marina. There are museums, galleries, restaurants, hotel and a theatre.*

Get more information about tourism and leisure activities in Swansea from this website: www.swansea.gov.uk/tourism

2.5 Tectonic processes

Is it safe to live in Zafferana?

This is the view that people in the village of Zafferana got on 22 July 2001 when Mount Etna erupted again (photo A). Etna, on the island of Sicily in southern Italy, is Europe's most active volcano, and every few years it sparks into life. This time it sent clouds of ash billowing high into the atmosphere and lava rolling down the mountainside. The lava was heading towards the Valle del Bove, the valley that lies directly above Zafferana (map B).

B 1:50,000 map extract of Mount Etna

A Mount Etna erupts in July 2001

Activities

1 Look at map B.
  a) Find the top of Mount Etna. How high is the volcano?
  b) Find Zafferana. How high is the village?
  c) Explain why Zafferana could be in danger from lava coming from Etna.

2 a) Measure the straight-line distance from Etna's main crater to Zafferana, using the scale line. How far is it to the nearest kilometre?
  b) Lava from Etna flows at about 2 metres per minute, or 120 m/hour. How long would it take to reach Zafferana?

  c) From your knowledge of volcanoes, what is likely to happen to the lava before it reaches the village?

3 Find evidence of previous lava flows on the map.
  a) In which years has Etna erupted in the past?
  b) When did the lava get closest to Zafferana? How far from the village did it stop?
  c) In that year, the Italian army used controlled explosions to divert the lava from the village. What could have happened if it reached the village?

4 Now answer the question: Is it safe to live in Zafferana? Give reasons for your answer.

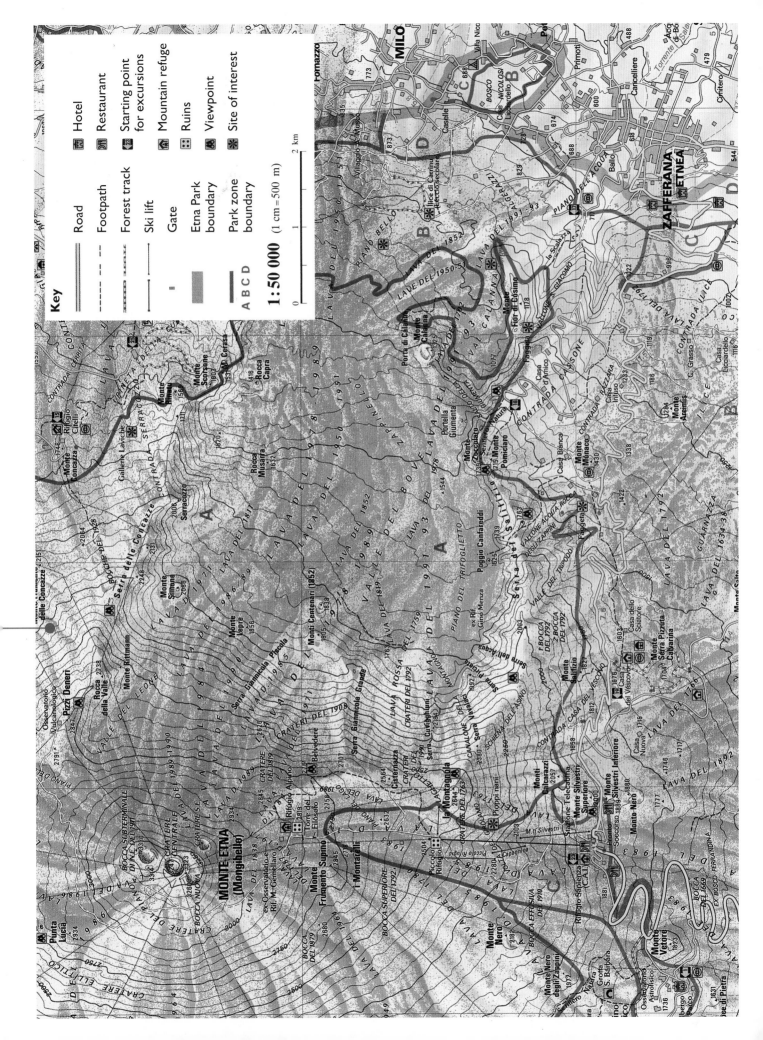

Key

	Road	🏨	Hotel
	Footpath	🍴	Restaurant
	Forest track	🚩	Starting point for excursions
	Ski lift	🏠	Mountain refuge
•	Gate	▦	Ruins
	Etna Park boundary	🔭	Viewpoint
A B C D	Park zone boundary	✳	Site of interest

1:50 000

0 — 1 — 2 km (1 cm = 500 m)

Why live so close to a volcano?

If volcanoes are dangerous, then why do people live so close? Satellite photo C gives us some clues. The colours on the photo show how land is used around Mount Etna. Find Zafferana and notice the land use pattern in the photo.

C | Satellite photo of Mount Etna

● Zafferana

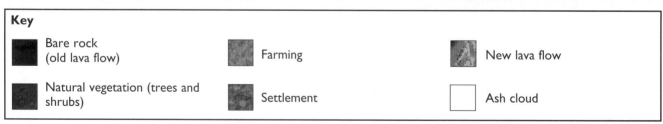

Key

Bare rock
(old lava flow)

Farming

New lava flow

Natural vegetation (trees and shrubs)

Settlement

Ash cloud

Activities

1 Look at photo C.
   Write a paragraph to describe the land use pattern on Mount Etna. You could start like this:

   Around the centre of the volcano is bare rock. Further from the centre there is.........

2 Answer these questions to explain the land use pattern in the photo.

a) Where is the main crater of the volcano? How can you tell from the photo?

b) Why is bare rock found close to the centre?

c) Why is natural vegetation found further from the centre?

d) Why do you think people live so close to the volcano?

D In 1992, lava from an eruption on Mount Etna almost reached Zafferana

Volcanic ash and lava is rich in minerals and, over time, it weathers to produce **fertile** soil. For centuries, people in Zafferana have used the soil to grow crops like olives and grapes. These days the village also has many hotels and a thriving tourist industry. People come from all over the world to visit Mount Etna.

E Growing vines on Etna

Assignment

You are going to draw a labelled cross-section of Mount Etna to show the land use zones and the position of Zafferana.

1 Look at map B. Place a strip of paper on the line from the top of Mount Etna to the centre of Zafferana. This is the line of your cross-section.

2 Draw a line the same length on a sheet of paper. This is the horizontal axis for your section.

3 Draw two vertical axes up to 3,300 metres (the height of Etna). Choose a suitable scale, so that the lines fit on your paper.

4 Draw your cross-section, following the instructions on page 17.

5 Mark the Etna Park boundary and the park zone boundaries on your cross-section. Then annotate these land use zones onto your cross-section:

- Zone A   Bare rock (old lava flow)
- Zone B   Natural vegetation (trees and shrubs)
- Zone D   Farmland (mainly olives and grapes)

Finally, mark the position of Zafferana.

2.6   *Environmental issues*

What can you do in a national park?

A **national park** is a large area of beautiful countryside, protected by law, so that it can be enjoyed by people, both now and in the future. National parks cover about 10% of the land area in England and Wales.

The Peak District National Park lies at the southern end of the Pennines in the middle of England. About 40,000 people live there, but it gets up to 30 million visitors a year. They get up to all sorts of activities!

A   Activities in the Peak District National Park

Hang gliding…you need a steep hill to take off

Activities

1 Here is a list of activities that you could do in the Peak District National Park:

> walking   cycling   climbing   caving
> pony trekking   sailing   fishing
> birdwatching   hang gliding   water skiing
> picnicking   exploring old villages
> visiting historic buildings

a) Find a place on map B where you could do each activity, and give a six-figure grid reference. For example, walking at 090860

b) For each activity, explain why this would be a good place to do it. For example, 090860 would be a good place to walk, because it is on the Pennine Way, a long-distance footpath.

2 Plan a one-week activity holiday in the area shown on map B. Think about:

a) how you would get there

b) where you would stay

c) what activities you would do each day

d) how you would get around.

You will get bonus points if you can plan a **sustainable**, or environmentally-friendly holiday. This means that you should make less use of cars, use public transport and walk or cycle instead. The activities you do should not harm the environment.

Caving…you need a big hole in the ground!

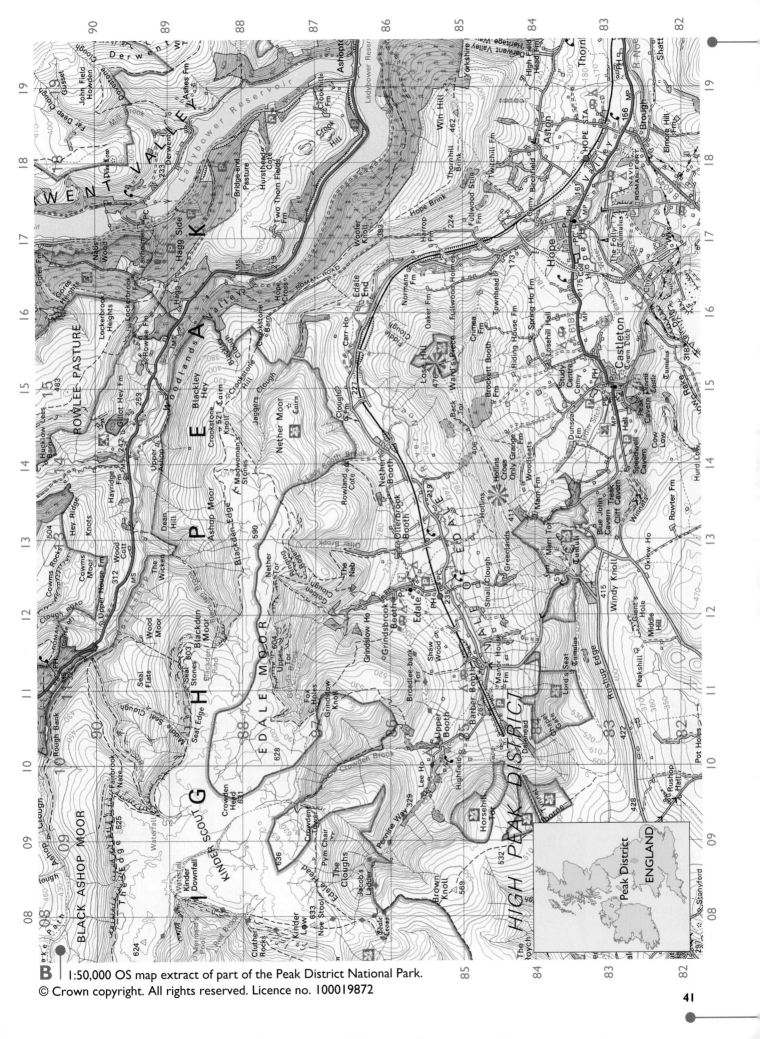

How should the park be managed?

Each national park is managed by a **National Park Authority (NPA)**. The NPA's main aims are:

- **conservation** to conserve the natural beauty of the park
- **recreation** to provide opportunities for visitors to enjoy the park
- **supporting the local economy** to meet the needs of local people for homes and jobs.

The Peak National Park Authority tries to balance the needs of different activities and the natural environment in the park. Sometimes these needs benefit each other (e.g. tourism and local business), but sometimes they can be in **conflict**. You can see the problems in photo C.

Local businesses need visitors to buy their goods and services. But, it's hard to cope when they all arrive at once.

Local residents have their lives disrupted by so many visitors. They complain that most of the local shops are for tourists – not for them.

C A busy weekend in Castleton, a village in the Peak District National Park

Farmers don't like visitors walking across their land. They get very upset if visitors drop litter or leave gates open when they should be shut.

Visitors come to the national park to enjoy the peace and quiet. But on sunny summer weekends there's not much of that!

Activities

1 Complete a matrix like this to show how different activities in the national park can conflict or benefit each other.

   a) For each box in the matrix, mark a '✗' where you think they conflict, mark a '✓' where you think the activities benefit each other, or a '0' where you don't think that they affect each other. For example, tourism and local business benefit each other, so mark a '✓' in the box.

   b) For each conflict that you identified, find a place on map B where you think the conflict is likely to happen. Give a grid reference for each one. For example, tourism and residents could conflict in the village of Castleton at 150830.

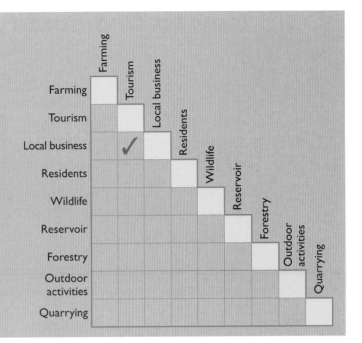

Assignment

You are going to make a management plan for the area of the Peak District National Park shown on map B.

To help to manage the area and to reduce any conflicts, divide it into zones. There are four types of zone you can use:

- **Wilderness zone** An area where natural vegetation and wildlife is conserved. A few people walk here to get away from it all.
- **Recreation zone** An area with many opportunities for outdoor activities. These can be on water, on land, or even underground.
- **Economic zone** An area where the main focus is on economic activities like farming, forestry or quarrying, and less on recreation.
- **Honeypot** A small area with shops and/or tourist attractions which attracts large numbers of visitors (and so keeps the numbers down in other zones).

1 Make a large copy of this grid. The numbers are the same as the grid lines on map B.

2 Now, divide the whole area on the grid into zones using map B and the information on the left. Draw the zones onto the grid and label them. You should have at least one example of each type of zone, but you can have more than that if you want.

3 Shade the zones on your map using four different colours, and give it a key.

4 Write a short report to explain why you divided the area into these zones. Refer to the features on map B that helped you to decide.

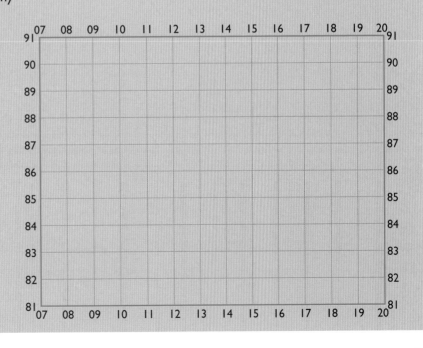

Key to Ordnance Survey 1:50,000 maps

Communications

ROADS AND PATHS | STRASSEN UND WEGE | VOIES DE COMMUNICATION

Not necessarily rights of way

Service area / Junction number

M 1   Elevated / En Viaduc / Erhöht

Motorway (dual carriageway)
Autoroute (chaussées séparées) avec aire de service et échangeur numeroté
Autobahn (zweibahnig) mit Servicestation und Anschlusstelle
Nummer der Anschlusstelle

Motorway under construction
Autoroute en construction
Autobahn im Bau

A 470   Unfenced / Dual carriageway / Sans clôture / Chaussées separées / Zweibahnige Strasse

Primary Route
Itinéraire principal
Fernstrasse

Primary route under construction
Itinéraire principal en construction
Fernstrasse im Bau

A 493   Main road
Route principale
Hauptstrasse

Main road under construction
Route principale en construction
Hauptstrasse im Bau

B 4518   Nicht eingezäunt

B 885   Secondary road
Route secondaire
Nebenstrasse

A 855   Narrow road with passing places
Route étroite avec voies de dépassement
Enge Strasse mit Ausweichstelle

Road generally more than 4m wide
Route généralement de plus de 4m de largeur
Strasse, im allg.über 4m breit

Road generally less than 4m wide
Route généralement de moins de 4m de largeur
Strasse, im allg.unter 4m breit

Other road, drive or track
Autre route, allée ou sentier
Sonstige Strasse, Zufahrt oder Feldweg

Footbridge / Passerelle / Fussgängerbrücke

Path / Sentier / Fussweg

Gradient : steeper than 20% (1 in 5)   14% to 20% (1 in 7 to 1 in 5)
Pente : Supérieure à 20% (1 pour 5)   14% à 20% (1 pour 7 à 1 pour 5)
Steigung über 20%   14% bis 20%

Bridge / Pont / Brücke

Gates / Barrières / Schranken

Road tunnel / Tunnel routier / Strassentunnel

Ferry (passenger) / Bac pour piétons / Personenfähre

Ferry (vehicle) / Bac pour véhicules / Autofähre

Ferry P   Ferry V

PRIMARY ROUTES

These form a network of recommended through routes which complement the motorway system

PUBLIC RIGHTS OF WAY | DROIT DE PASSAGE PUBLIC | ÖFFENTLICHE WEGERECHTE

Footpath
Road used as a public path
Bridleway
Byway open to all traffic

Public rights of way shown on this map have been taken from local authority definitive maps and later amendments. The map includes changes notified to Ordnance Survey by 1st August 1997.
The symbols show the defined route so far as the scale of mapping will allow.
Rights of way are not shown on maps of Scotland

OTHER PUBLIC ACCESS | AUTRES ACCES PUBLICS | ANDERE ÖFFENTLICHE WEGE

● National Trail, European Long Distance Path, Long Distance Route, selected Recreational Routes

● ● ● Other route with public access (not normally shown in urban areas)

The exact nature of the rights on these routes and the existence of any restrictions may be checked with the local highway authority. Alignments are based on the best information available. These routes are not shown on maps of Scotland

National/Regional Cycle Network

● Surfaced cycle route

Danger Area   Firing and Test Ranges in the area. Danger! Observe warning notices.
Champs de tir et d'essai. Danger! Se conformer aux avertissements.
Schiess und Erprobungsgelände. Gefahr! Warnschilder beachten.

4   National Cycle Network number
8   Regional Cycle Network number

RAILWAYS | CHEMINS DE FER | EISENBAHNEN

Track multiple or single
Track under construction
Light rapid transit system, narrow gauge or tramway
Bridges, Footbridge
Tunnel

a   Station, (a) principal
Siding
Light rapid transit system station
LC   Level crossing
Viaduct

General Information

LAND FEATURES

Electricity transmission line (pylons shown at standard spacing)

Pipe line (arrow indicates direction of flow)

Buildings

Public building (selected)

Bus or coach station

Place of Worship   with tower / with spire, minaret or dome / without such additions

Chimney or tower

Glass Structure

Heliport

Triangulation pillar

Mast

Wind pump/wind generator

Windmill with or without sails

Graticule intersection at 5' intervals

Cutting, embankment

Quarry

Spoil heap, refuse tip or dump

Coniferous wood

Non-coniferous wood

Mixed wood

Orchard

Park or ornamental ground

Forestry Commission access land

National Trust-always open

National Trust-limited access, observe local signs

National Trust for Scotland

Rights of way are liable to change and may not be clearly defined on the ground. Please check with the relevant local authority for the latest information

The representation on this map of any other road, track or path is no evidence of the existence of a right of way

Tourist Information

RENSEIGNEMENTS TOURISTIQUES / TOURISTENINFORMATION

TOURIST INFORMATION

⌂	Camp site / Terrain de camping / Campingplatz
	Caravan site / Terrain pour caravanes / Wohnwagenplatz
✳	Garden / Jardin / Garten
⌐	Golf course or links / Terrain de golf / Golfplatz
🛈 *i*	Information centre, all year / seasonal / Office de tourisme, ouvert toute l'année / en saison / Informationsbüro, ganzjährig / saisonal
	Nature reserve / Réserve naturelle / Naturschutzgebiet
P P&R	Parking / Park and ride, all year / seasonal / Parking / Parking et navette, ouvert toute l'année / en saison / Parkplatz / Park & Ride, ganzjährig / saisonal
✕	Picnic site / Emplacement de pique-nique / Picknickplatz
○ PC	Public convenience (in rural areas) / Toilettes (à la campagne) / Öffentliche Toilette (in ländlichen Gebieten)
	Selected places of tourist interest / Endroits d'un intérêt touristique particulier / Ausgewählter Platz von touristischem Interesse
☏ ✆	Telephone, public / motoring organisation / Téléphone, public / associations automobiles / Telefon, öffentlich / automobilklub
❀	Viewpoint / Point de vue / Aussichtspunkt
V	Visitor centre / Centre pour visiteurs / Besucherzentrum
👣 🚶	Walks / Trails / Promenades / Wanderwege
▲	Youth hostel / Auberge de jeunesse / Jugendherberge

Technical Information

NORTH POINTS

Difference of true north from grid north at sheet corners

NW corner	NE corner
1° 03' (19 mils) E	0° 33' (10 mils) E
SW corner	SE corner
1° 02' (18 mils) E	0° 32' (10 mils) E

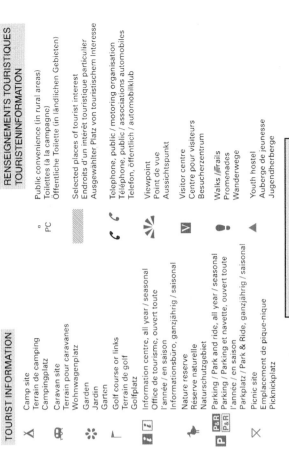

True North
Grid North
Magnetic North
Diagrammatic only

Magnetic north varies with place and time. The direction for the centre of the sheet is estimated at 3° 19' (59 mils) west of grid north for July 2004.

Annual change is about13' (4 mils) east

Magnetic data supplied by the British Geological Survey

Base map constructed on Transverse Mercator Projection, Airy Spheroid, OSGB (1936) Datum. Vertical datum mean sea level (Newlyn)

To plot the average direction of magnetic north join the point circled on the point on the south edge of the sheet to the point on the protractor scale on the north edge at the angle estimated for the current year

INCIDENCE OF ADJOINING SHEETS

The red figures give the grid values of the adjoining sheet edges. The blue letters identify the 100 000 metre square.

HOW TO GIVE A NATIONAL GRID REFERENCE TO NEAREST 100 METRES

SAMPLE POINT: Goodcroft

1. Read letters identifying 100 000 metre square in which the point lies ....NY

2. FIRST QUOTE EASTINGS
Locate first VERTICAL grid line to LEFT of point and read LARGE figures labelling the line either in the top or bottom margin or on the line itself. 53
Estimate tenths from grid line to point. 4

3. AND THEN QUOTE NORTHINGS
Locate first HORIZONTAL grid line BELOW point and read LARGE figures labelling the line either in the left margin or on the line itself. 16
Estimate tenths from grid line to point. 1

SAMPLE REFERENCE NY 534 161

For local referencing grid letters may be omitted

IGNORE the SMALLER figures of the grid number at the corner of the map. These are for finding the full coordinates. Use ONLY the LARGER figure of the grid number. EXAMPLE: 3 1 **7** 000m

---

BOUNDARIES

Administrative boundaries as at January 2002

＋ — ＋	National
＋ ┼ ＋	District
— ·· — · —	County, Unitary Authority, Metropolitan District or London Borough
	National Park

WATER FEATURES

Marsh or salting
Towpath
Aqueduct
Canal
Weir
Lake
Footbridge
Bridge
Lock
Ford
Normal tidal limit
Canal (dry)

Contour values in lakes are in metres

ABBREVIATIONS

CH	Clubhouse
MS	Milestone
PC	Public convenience (in rural area)
TH	Town Hall, Guildhall or equivalent
CG	Coastguard
P	Post office
MP	Milepost
PH	Public house

ARCHAEOLOGICAL AND HISTORICAL INFORMATION

✚	Site of monument
○	Stone monument
⚔	Battlefield (with date)
☆ ···	Visible earthwork
VILLA	Roman
Castle	Non-Roman

Information provided by English Heritage for England and the Royal Commissions on the Ancient and Historical Monuments for Scotland and Wales

HEIGHTS

——50—— Contours are at 10 metres vertical interval

144 Heights are to the nearest metre above mean sea level

Heights shown close to a triangulation pillar refer to the ground at the base of the pillar and not necessarily to the summit

ROCK FEATURES

Outcrop
Cliff
Scree

High water mark
Low water mark
Cliff
Slopes
Flat rock
Sand
Dunes
Mud
Shingle
△ Beacon
⌇ Lighthouse (in use)
⌇ Lighthouse (disused)

CONVERSION

METRES – FEET

1 metre = 3.2808 feet

1000	3000
900	
800	2500
700	2000
600	
500	1500
400	
300	1000
200	
100	500
0	0 Feet

Metres 0

15.24 metres = 50 feet

Index

Page numbers in *italic* refer to photographs, maps and drawings

aerial photos 2
agriculture, on slopes of Mount Etna 39, *39*
Ashford, Kent, growth of
    favourable location 26
    in the future 27
    government choice as a growth area 27
    housing *26*
Ashford, Kent OS map extract (1:25, 000)
    25

brownfield sites 27

central business district (CBD) 24
cliffs *13*
    Seven Sisters *14*
climate change, and increased flooding 23
compass, points of 6, *6*
contour lines 12, *13*, 14, 30
cross-section drawing 17, *17*, 30
Cuckmere River Valley *5*, *6*, *12*

decision-making using a map 18, *19*
direction 6–7
distance 10–11
    measured by road 11, *11*
    straight line (as the crow flies) 10, *10*

economic zone 43
Environment Agency, and flood defences
    23
environmental issues 40–43
Etna, Mount, Sicily, Italy see Mount Etna

flood defences, Shrewsbury *21*
flood risk, in Shrewsbury 20, *20*, *21*, 22
    reduced 23, *23*
floodplains 22
    floodplain map, Shrewsbury *22*
floods/flooding 20
    Shrewsbury, Autumn 2000 22

greenfield sites 27
grid references *21*
    four-figure 8
    six-figure 9

Helvellyn OS map extract (1:50,000) *28*
hill contours *13*
honeypot sites 43

industrial estates 33
industrial location
    Ashford 26
    Lower Swansea Valley 34

Lake District, work of the water cycle 28, *29*
land use 38–39, *38*, *39*
lava 36, 39
Lower Swansea Valley, changes in 32–35
1952, not a good industrial location 34
    coal mining declined 32
    loss of metal smelting industry 32
    new industrial estates 33
    regeneration since 1980 35
    tourism 33, 35
Lower Swansea Valley OS map extract
    1:50,000 (2003) *33*
    1:66,000 (1952) *32*

management plan, Peak District National
    Park 43
maps 2–3
    for decision making 18
metal smelting 32
Mount Etna
    map extract (1:50,000) *37*
    reasons for living close to 38–39
    satellite photo *38*
    volcanic eruptions 36, *36*
    Zafferana, a safe place to live? 36
mountains
    planning to walk in safety 30–31
    planning a walk to top of Helvellyn 31

National Park Authority, aims 42
national parks 40
    management 42
new homes in the UK, options 27

Ordnance Survey maps
    1:50,000 maps, key to *44–45*
    key on the sheet 4

Peak District National Park
    OS map extract (1:50,000) *41*
    possible activities 40, *40*
Peak Park National Authority, balancing
    various needs 42
physical features, symbols for *5*
plans 3

recreation zone 43
regneration, Lower Swansea Valley 35

relief 14, *14*, *15*
    spot heights and contours 12–13
rural areas, symbols *5*

satellite photos *38*
scale 10–11, 15
Seaford *5*
    possible sites for superstore *19*
    walking along the Vanguard Way 15, *15*
Seaford OS map 2–19
    four-figure grid references 8
    six-figure grid references 9
settlements 24–27
    near rivers 20
Severn, River 20, *21*, 22–23
Shrewsbury
    flood risk 30, *31*
    flooding, Autumn 2000 22
    floodplain map for *22*
    reasons behind its location 20
Shrewsbury OS map extract (1:25,000) *20*
sketch map drawing 16, *21*
    labelling/annotation *16*, 29
slope, and contours 13
soils, fertile, from weathering of volcanic
    ash and lava 39
South Downs Way 7
spot heights 12
superstores, finding sites for 18
symbols 4, *4*, *5*, *14*, 44–45

tourism 33, 35
transport connections, favour Ashford's
    growth 26
triangulation pillars 12

urban areas, symbols *5*
urban land use model 24

valley contours *13*
Vanguard Way 15, *15*
    drawing a cross-section *17*
volcanic ash 36, 39
volcanoes
    Mount Etna 36
    reasons for living close to 38–39

water cycle, in the Lake District 28, *29*
weather forecasts, and safety on mountain
    walks 31
wilderness zone 43